STUDENT GUIDE

THE LANGUAGE of NUMBERS

INVENTING AND COMPARING NUMBER SYSTEMS

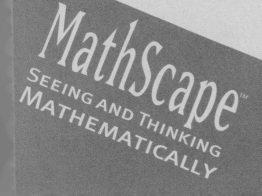

MathScape
SEEING AND THINKING
MATHEMATICALLY

How is our current number system like an ancient number system?

THE LANGUAGE
OF
NUMBERS

PHASE**ONE**
Mystery Device™

Our everyday number system is one of humanity's greatest inventions. With just a set of ten simple digits, we can represent any amount from 1 to a googol (1 followed by 100 zeros) and beyond. But what if you had to create a new system? In Phase One, you will investigate the properties of a number system. To do this, you will be using a Mystery Device to invent a new system.

PHASE**TWO**
Chinese Abacus

The Chinese abacus is an ancient device that is still used today. You will use the abacus to solve problems such as: What 3-digit number can I make with exactly three beads? You will compare place value in our system to place value on the abacus. This will help you to better understand our number system.

PHASE**THREE**
Number Power

In this phase, you will test your number power in games. This will help you see why our number system is so amazing. You will explore systems in which place values use powers of numbers other than 10. You will travel back in time to decode an ancient number system. Finally, you will apply what you have learned to create the ideal number system.

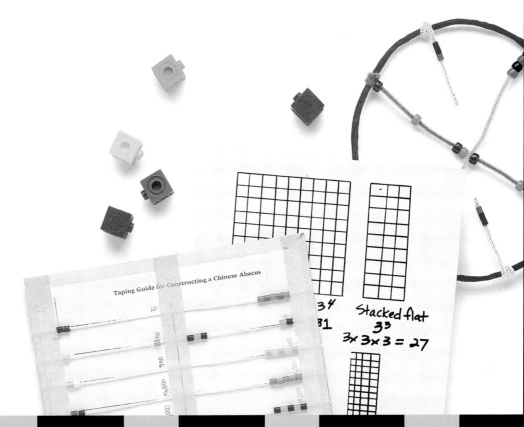

PHASE ONE

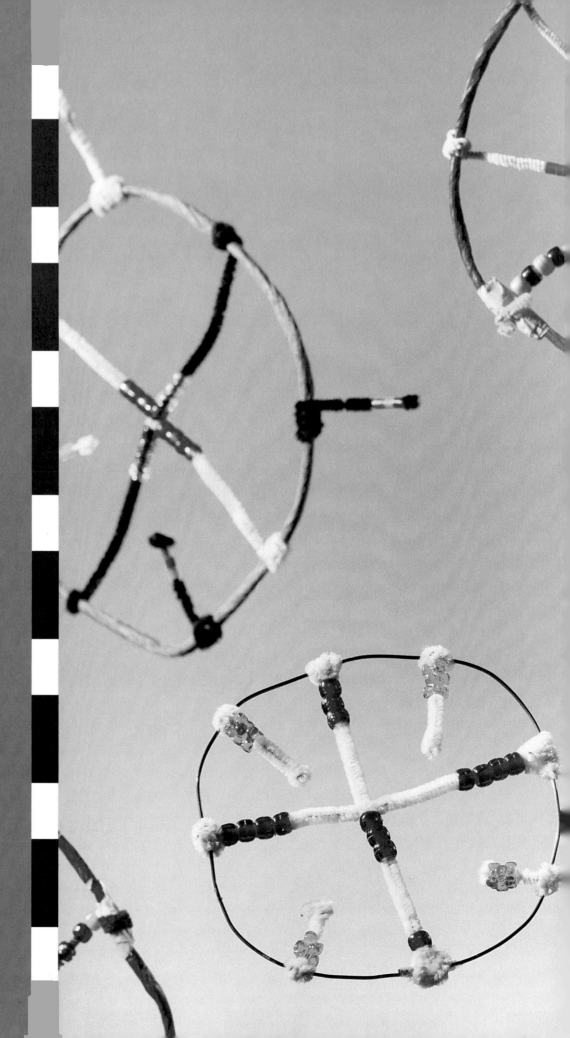

Imagine a mysterious number-making device has been discovered. The device does not work with our everyday number system. Only you can unlock its secrets.

What do computer programmers and experts in cracking codes have in common? For people in these careers, understanding number systems is an important skill. Can you think of other careers in which number systems are important?

Mystery Device

WHAT'S THE MATH?

Investigations in this section focus on:

PROPERTIES of NUMBER SYSTEMS

- Identifying different properties of a number system

- Analyzing a new number system, and comparing it to our everyday number system

- Making connections between number words and a number system's rules

- Describing a number system as having symbols, rules, and properties

NUMBER COMPOSITION

- Using expanded notation to show how numbers are made in different systems

- Writing arithmetic expressions for number words

- Recognizing that the same number can be written in different ways

- Finding arithmetic patterns in number words

MathScape Online
mathscape1.com/self_check_quiz

 # Inventing a **Mystery** Device System

REPRESENTING
NUMBERS IN
DIFFERENT WAYS

Some pipe cleaners and beads are all you need to make your own Mystery Device. You will use it to invent your own system for making numbers. Can you make rules so that others will be able to use your system?

Create a Mystery Device

What would you need to invent a number system?

Use the Mystery Device Assembly page to make your own Mystery Device. Your Mystery Device will look like this when it is complete. Make sure that the short "arms" can be turned outward as well as inward.

Make Numbers Using the Mystery Device

Find a way to make all the numbers between 0 and 120 on your Mystery Device. See if you can create one set of rules for making all of the numbers on the Mystery Device. Use the following questions to test your new system.

How can we use a Mystery Device to represent numbers?

- How does my system use the beads to make a number? Does the size of a bead or the position of a bead or an arm make a difference?

- Does my system work for large numbers as well as small numbers? Do I have to change my rules to make any number?

- How could I explain to another person how to use my system? Would it make a difference which part of the device was at the top?

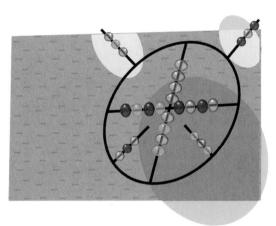

Describe the Invented Number System

Describe the rules for your Mystery Device number system. Use drawings, charts, words, or numbers. Explain your number system so that someone else could understand it and use it to make numbers.

1 Explain in words and drawings how you used your system to make each of the following numbers: 7, 24, 35, 50, 87, and 117.

2 Explain in words and pictures how you made the largest number it is possible to make in your system.

hot **words** | number system
number symbols
expanded notation

3 Explain how you can use expanded notation to show how you composed a number in your system.

Homework
page 80

How is your system different from the other systems in the classroom?

Comparing Mystery Device Systems

What are the "building blocks" of a number system? To find out, you will make different numbers on the Mystery Device. You will invent your own way to record them. See how the building blocks of your Mystery Device system compare to those in our number system.

How can you use expanded notation to show how you made a number in your system?

Explore Expanded Notation

Use your Mystery Device to make these numbers. Come up with a system of expanded notation to show how you made each number on the Mystery Device.

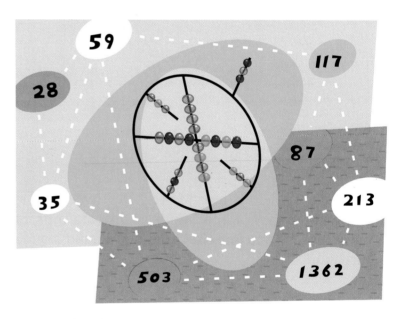

Why is it important for the class to agree on one method of expanded notation?

Investigate the Building Blocks of Number Systems

Figure out the different numbers you can make on your Mystery Device using 3 beads. The beads you use can change from number to number, but you must use only 3 for each number. Keep a record of your work using expanded notation.

What numbers can you make using 3 beads?

- What do you think is the least number you can make with only 3 beads? the largest?

- If you could use any number of beads, could you make a number in more than one way?

- Do you think there are numbers that can be made in only one way?

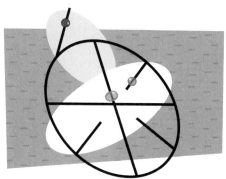

$$20 + 4 + 1 = 25$$

Compare Number Systems

Answer the five questions below to compare your Mystery Device system to our number system. Then make up at least three of your own questions for comparing number systems.

1 Can you make a 3-bead number that can be written with exactly 3 digits in our number system?

2 Can you make a 3-bead number with more than 3 digits?

3 Can you make a 3-bead number with fewer than 3 digits?

4 What are the building blocks of the Mystery Device system?

5 What are the building blocks of our number system?

hot **words** | arithmetic expression
expanded notation

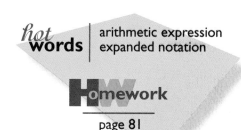

Homework
———
page 81

3 Number Words in Many Languages

FINDING
ARITHMETIC
PATTERNS

Patterns in the number words of other languages can help you see how numbers can be made. Here you will search for patterns in number words from different languages. This will help you understand the arithmetic behind some English number words.

Find Patterns in Number Words from Fulfulde

What can you learn about number systems from looking at number words in many languages?

Look at the Fulfulde words for 1–100. Figure out how each of the Fulfulde number words describes how a number is made. Beside each number word, write an arithmetic expression that shows the building blocks for that number. The *e* shows up in many of the number words. What do you think *e* means?

Number Words in Fulfulde (Northern Nigeria)			
1	go'o	15	sappo e joyi
2	didi	16	sappo e joyi e go'o
3	tati	17	sappo e joyi e didi
4	nayi	18	sappo e joyi e tati
5	joyi	19	sappo e joyi e nayi
6	joyi e go'o	20	noogas
7	joyi e didi	30	chappan e tati
8	joyi e tati	40	chappan e nayi
9	joyi e nayi	50	chappan e joyi
10	sappo	60	chappan e joyi e go'o
11	sappo e go'o	70	chappan e joyi e didi
12	sappo e didi	80	chappan e joyi e tati
13	sappo e tati	90	chappan e joyi e nayi
14	sappo e nayi	100	teemerre

How are Fulfulde number words similar to English number words?

Decode Number Words from Another Language

Work as a group to complete each of the following steps. Decode the number words using either a Hawaiian, Mayan, or Gaelic number words chart.

1 Write an arithmetic expression for each word on the number chart.

2 Predict what the number words for 120, 170, 200, and 500 would be in the language that you selected.

3 Write an arithmetic expression for each new number word and explain how you created the new number word.

What do the different languages have in common in the way they make number words?

How can you use arithmetic expressions to compare number words in different languages?

Create a Mystery Device Language

Invent a Mystery Device language that follows the rules of at least one number system you have decoded. Use the chart below as an example. Make up number words for 1–10 in your own Mystery Device language.

1 Using your new Mystery Device language, try to create words for the numbers 25, 43, 79, and 112. The words should describe how these numbers would be made on your Mystery Device.

2 Write an arithmetic expression to show how you made each number.

1	en
2	sessi
3	soma
4	vinta
5	tilo
6	chak
7	bela
8	jor
9	drona
10	winta

hot **words** | multiple pattern

Homework

page 82

Examining Alisha's System

ANALYZING A NEW NUMBER SYSTEM

How well does the number system invented by Alisha work? You will use what you have learned to analyze Alisha's number system and language. See if Alisha's system works well enough to become the Official Mystery Device System!

Analyze Alisha's Mystery Device System

How does Alisha's system work?

Use the chart below to figure out Alisha's system. As you answer each question, make a drawing and write the arithmetic expression next to it. Only show the beads you use in each drawing.

1 How would you make 25 in Alisha's system, using the least number of beads?

2 Choose two other numbers between 30 and 100 that are not on the chart. Make them, using the least number of beads.

3 What is the greatest number you can make?

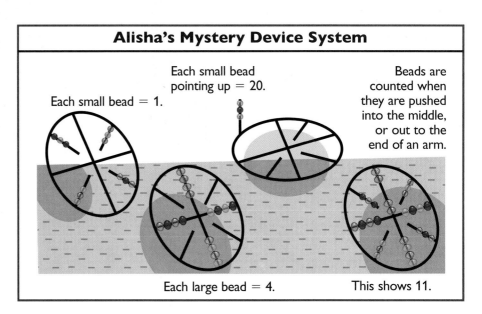

Alisha's Mystery Device System

Each small bead = 1.

Each small bead pointing up = 20.

Beads are counted when they are pushed into the middle, or out to the end of an arm.

Each large bead = 4.

This shows 11.

Make Number Words in Alisha's System

How is Alisha's system like our number system?

Alisha also made up number words to go with her Mystery Device system. They are shown in the table. Answer the questions below to figure out how Alisha's system works.

1 Tell what number each number word represents and write the arithmetic expression.

   a. soma, sim-vinta, en   **b.** set-soma, vintasim

   c. sim-soma, set   **d.** vinta-soma, set-vinta

   e. vintaen-soma, set-vinta, sim

2 Write the word in Alisha's system for 39, 95, and 122.

1	en	11	set-vinta, sim	30	soma, set-vinta, set
2	set	12	sim-vinta	40	set-soma
3	sim	13	sim-vinta, en	50	set-soma, set-vinta, set
4	vinta	14	sim-vinta, set	60	sim-soma
5	vintaen	15	sim-vinta, sim	70	sim-soma, set-vinta, set
6	vintaset	16	vinta-vinta	80	vinta-soma
7	vintasim	17	vinta-vinta, en	90	vinta-soma, set-vinta, set
8	set-vinta	18	vinta-vinta, set	100	vintaen-soma
9	set-vinta, en	19	vinta-vinta, sim		
10	set-vinta, set	20	soma		

Evaluate Number Systems

Use these questions to evaluate your Mystery Device system and Alisha's system. Decide which one should become the Official Mystery Device System. Explain your reasons.

- What are two things that an Official Mystery Device System would need to make it a good number system?

- Which of the two things you just described does Alisha's system have? Which of them does your system have? Give examples to show what you mean.

- What is one way you would improve your system to make it the Official Mystery Device System?

hot **words** | rule
arithmetic expression

Homework
page 83

THE LANGUAGE OF NUMBERS • LESSON 4   **59**

PHASE TWO

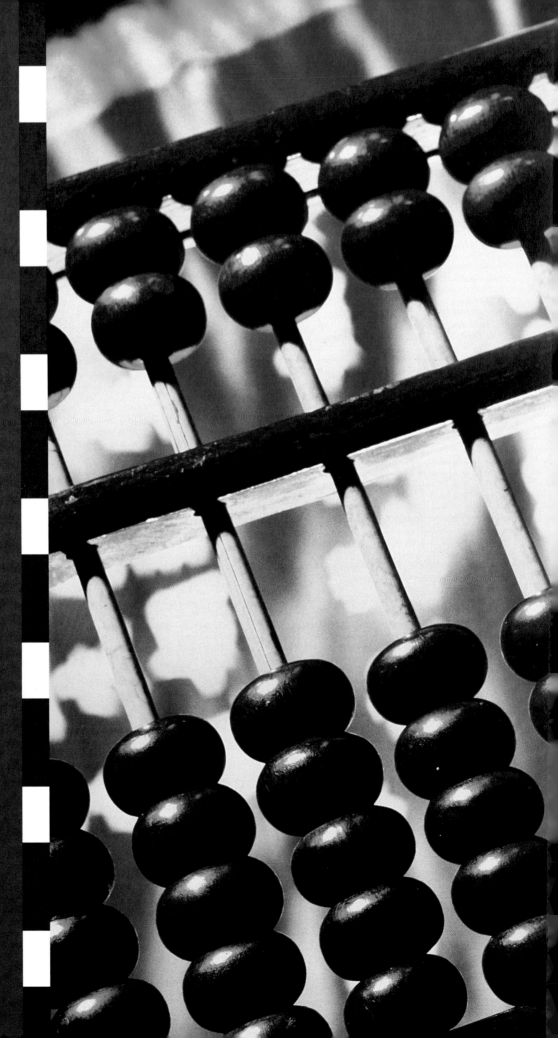

This counting instrument is called *choreb* in Armenian. In Japanese, it is a *soroban,* and the Turks know it as the *coulba.* The Chinese call it a *suan pan* or *sangi.* Most of us know it by the Latin name *abacus.*

Different forms of the abacus have developed in different cultures around the world. Many are still widely used today. You may be familiar with the Chinese, Japanese, Russian, or other abaci. The abacus helps us to see how place value works in a number system.

Chinese Abacus

WHAT'S THE MATH?

Investigations in this section focus on:

PROPERTIES of NUMBER SYSTEMS

- Representing and constructing numbers in a different number system

- Investigating and contrasting properties of number systems

- Understanding the use and function of place value in number systems

NUMBER COMPOSITION

- Understanding the connection between trading and place value in number systems

- Recognizing patterns in representing large and small numbers in a place-value system

- Understanding the role of zero as a place holder in our own place-value number system

MathScape Online

mathscape1.com/self_check_quiz

Exploring the Chinese Abacus

As on the Mystery Device, you move beads to show numbers on the Chinese abacus. But you will find that in other ways the abacus is more like our system than the Mystery Device. Can you find the ways that the abacus system is like our system?

Make Numbers on a Chinese Abacus

How does an abacus make numbers?

The columns on this abacus are labeled so that you can see the values. See if you can follow the Chinese abacus rules to make these numbers: 258; 5,370; and 20,857.

Chinese Abacus Rules

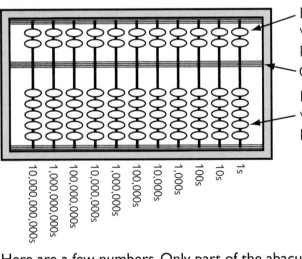

Beads above worth 5 times place value

Crossbar

Beads below worth 1 times place value

10,000,000,000s
1,000,000,000s
100,000,000s
10,000,000s
1,000,000s
100,000s
10,000s
1,000s
100s
10s
1s

- Each column on the Chinese abacus has a different value.

- A crossbar separates the abacus into top and bottom sections.

- Each bead above the crossbar is worth **5** times the value of the column if pushed toward the crossbar.

- Each bead below the crossbar is worth **1** times the value of the column if pushed toward the crossbar.

- A column shows 0 when all the beads in the column are pushed away from the crossbar.

Here are a few numbers. Only part of the abacus is shown.

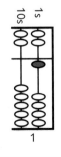
1

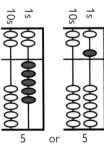
5 or 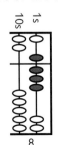 5

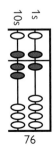
8

76

Investigate the Chinese Abacus

For each investigation below, explore different ways to make numbers on your abacus. Use both a drawing and an arithmetic notation to show how you made each number.

How is place value on a Chinese abacus like place value in our system? How is it different?

1 Make each of these numbers on the abacus in at least two different ways.

    a. 25   **b.** 92   **c.** 1,342   **d.** 1,000,572

2 Use any 3 beads to find these numbers. You can use different beads for each number, but use exactly three beads.

    a. the greatest number you can make

    b. the least number you can make

3 Find some numbers we write in our system with 3 digits that can be made with exactly 3 beads.

    a. Find at least five different numbers that you can make with 3 beads.

    b. Make at least one number using only the first two columns of the abacus.

Define Place Value

Our number system uses place value. Each column has a value, and 0 is used as a place holder, so that 3 means 3 ones, 30 means 3 tens, 300 means 3 hundreds, and so on. How is the use of place value on the Chinese abacus like its use in our system? How is it different?

- Write a definition for place value that works for both the Chinese abacus and our system.

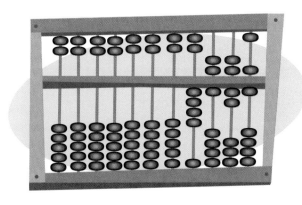

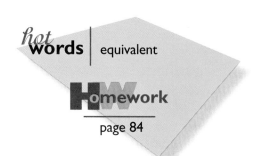

hot **words** | equivalent

Homework

page 84

 # How Close Can You Get?

TRADING AND
PLACE VALUE

How close can you get to a target number using a given number of beads? In this game you will explore trading among places on the Chinese abacus. You can use what you discover in playing the game to compare the Chinese abacus to our system.

How close can you get to 6,075 using exactly 14 beads?

Investigate Trading Relationships

Follow these steps to play the game How Close Can You Get. Try to get as close as possible to the target number using the given number of beads. If you can't make the number exactly, get as close as you can.

How Close Can You Get? Game Rules

1. One player picks a target number between 1,000 and 9,999.

2. Another player picks a number of beads, from 7 to 16. You must use the exact number of beads selected.

3. All players write down the group's challenge for the round: How close can you get to _____ using exactly _____ beads?

4. When all players have made a number, compare answers. The player or players who come closest to the target number score one point.

5. Continue playing, with different players picking the target number and number of beads. When someone reaches 10 points, the game is over.

When do we use trading in our number system?

Solve the Mystery Number Puzzles

Here are four mystery number puzzles. To solve each puzzle, you need to figure out which part of the abacus might be shown and give at least one number that the beads might make. Use drawings and expanded notation to show your answers. Beware! One of the mystery number puzzles is impossible to solve, and some puzzles can be solved in more than one way.

What trades can you make on the Chinese abacus to make both 5,225 and 5,225,000 with only 12 beads?

Puzzle A

What number might be shown?

Clues:
- All the beads used to make the number are shown.
- One of the columns is the 10,000s column.
- The 10s column is not shown at all.

Puzzle B

What number might be shown?

Clues:
- All the beads used to make the number are shown.
- One of the columns is the 100s column.
- The number is between 100,000 and 10,000,000.

Puzzle C

What number might be shown when you add the missing bead?

Clues:
- All the columns used to make the number are shown.
- One bead is missing from the figure.
- When we write the target number in our system, there is a 1 in the 1,000,000s place.

Puzzle D

What number might be shown when you add the missing beads?

Clues:
- All the columns used to make the number are shown.
- There are two beads missing from the figure.
- Beads are used only in the top part of the abacus to make the number.
- When we write the target number in our system, the only 5 used is in the 1,000s place.

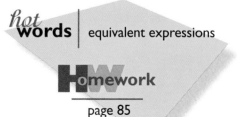

hot **words** | equivalent expressions

HW**omework**

page 85

7 Additive Systems

EXAMINING A
DIFFERENT KIND
OF SYSTEM

An additive system does not use place value. You simply add together the values of individual symbols to find the value of the number. For example, if △ equals 1 and □ equals 7, then □ □ △ equals 7 + 7 + 1 = 15. Do you think a number system like this would be easier or harder to use than our system?

What if place value was not used at all in a number system?

Investigate How Additive Systems Work

Figure out how your additive system works by making the following numbers. Record how you made each number on a chart. Write arithmetic expressions for the three greatest numbers.

1 Make the numbers 1 through 15.

2 Make the greatest number possible with three symbols.

3 Make five other numbers greater than 100.

Number in our System	Number in Additive System	Arithmetic Expression (for three greatest numbers only)
10	□ △ △ △	7 + 1 + 1 + 1 = 10

What are the patterns in your system?

Compare the Three Systems

In our system, three-digit numbers are always greater than two-digit numbers. For example, 113 has three digits. It is greater than 99, which has two digits. In the system you are investigating, are numbers that have three symbols always greater than numbers that have two symbols? Explain why or why not.

Improve the Number Systems

Improve your additive number system by making up a new symbol. The new symbol should make the system easier to use or improve it in some other way.

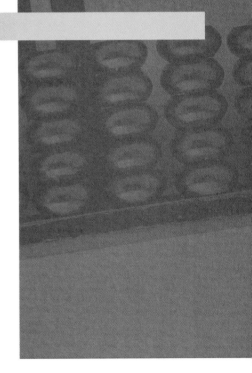

How does a new symbol improve the system?

1 Give the new symbol a value different from the other symbols in the system.

2 Make at least five numbers with your improved system. Write them on your recording sheet.

Analyze the Improved Additive Systems

Work with a partner who investigated a different system. See if you can figure out how each other's improved systems work. Compare the largest numbers you can make with three symbols and with four symbols. Answer the following questions.

- Which system lets you make greater numbers more easily? Why?

- Which patterns of multiples are easy to recognize in each system? Are they the same? Why or why not?

- Can you find other ways in which the two systems are alike?

- Can you find other ways in which they are different?

- In what ways are these number systems like our system?

hot **words** | additive system
Roman numerals

H**omework**

page 86

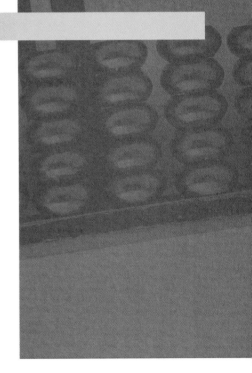

THE LANGUAGE OF NUMBERS • LESSON 7   **67**

The MD System

IDENTIFYING THE
PROPERTIES OF
NUMBER SYSTEMS

It's time to get out your Mystery Device and learn a new number system: the MD system. As you learn to use the MD system, you will investigate making numbers in more than one way. To do this, you will play How Close Can You Get.

Decode the MD Number System

How would you make 25 using the MD system?

This illustration shows how much the beads are worth in the MD system. The Mystery Device here shows 0—all of the beads are away from the center. To make numbers, you push beads toward the center of the Mystery Device. Always keep the short arms inside the hoop.

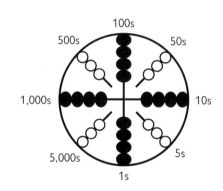

Create Multiple Representations of Numbers

Can you make numbers in more than one way in the MD system?

Use the MD system to make each of the following numbers on the Mystery Device.

1 Make a 4-digit number with 0 in one of the places.

2 Make a 3-digit number that can be made in at least two ways.

3 Make a number that can be shown in only one way.

Investigate Trading Relationships in the MD System

Jackie is playing How Close Can You Get? and she needs help. She has made 6,103 with 6 beads, but she doesn't know what to do next to get to 12 beads. Write a hint to Jackie using words, drawings, or arithmetic notation. Be careful not to give away the answer.

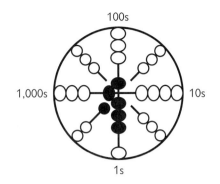

How close can you get to 6,103 with exactly 12 beads using the MD number system?

Compare the MD System to Other Systems

Make a chart like the one shown. Fill in the chart to compare the MD number system to your Mystery Device system, the abacus system, the additive system you investigated in Lesson 7, and our own number system.

How is the MD number system similar to each of the other systems?

Comparison Questions	MD System	My Mystery Device System	Chinese Abacus	Additive System Investigated	Our Number System
What are the kinds of symbols used to make numbers?					
What are the building blocks?					
Is there a limit on the highest number possible?					
Is there more than one way to show a number?					
How does the system use place value?					
How does the system use trading?					
How does the system use zero?					
What are the patterns in the system?					
How is the system additive?					

hot **words** | place value

Ho**mework**
page 87

PHASE THREE

Our system for representing numbers was developed over thousands of years. People from cultures all over the world have had a part in making it such a powerful tool for working with numbers.

Imagine that you have been asked to investigate different number systems to help improve our number system. How would you create a "new and improved" number system?

Number Power

WHAT'S THE MATH?

Investigations in this section focus on:

PROPERTIES of NUMBER SYSTEMS

- Understanding that number systems are efficient if every number can be represented in just one way and only a few symbols are used

- Describing in detail the features of our number system

- Identifying and describing a mathematically significant improvement to a number system

NUMBER COMPOSITION

- Writing an arithmetic expression using exponents

- Learning how to evaluate terms with exponents, including the use of 0 as an exponent

- Developing number sense with exponents

MathScape Online
mathscape1.com/self_check_quiz

Stacks and Flats

WRITING
ARITHMETIC
EXPRESSIONS USING
EXPONENTS

Our number system is a base 10 place-value system. You can better understand our system by exploring how numbers are shown in other bases. As you explore other bases, you will learn to use exponents to record the numbers you make.

Make Numbers in the Base 2 System

How can you show a number using the fewest base 2 pieces?

Make a set of pieces in a base 2 system. Use your base 2 pieces to build the numbers 15, 16, 17, 26, and 31. Write an arithmetic expression using exponents for each number. You may need to make more pieces to build some numbers.

1 Fill in your Stacks and Flats Recording Sheet to tell how many pieces you used to make each number. Write 0 for the pieces you do not use.

2 Write an arithmetic expression using exponents for each number you make.

How does the pattern of exponents in base 2 compare to the pattern of exponents in base 10?

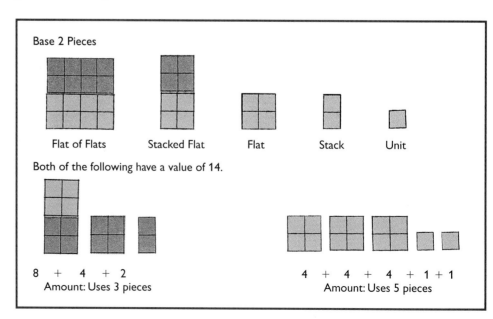

Base 2 Pieces

Flat of Flats   Stacked Flat   Flat   Stack   Unit

Both of the following have a value of 14.

8 + 4 + 2
Amount: Uses 3 pieces

4 + 4 + 4 + 1 + 1
Amount: Uses 5 pieces

Investigate a Base 3 or Base 4 Number System

Choose whether you will work in base 3 or base 4. Make a new set of pieces for the base you have chosen. Your set should include at least 5 units, 5 stacks, 5 flats, 3 stacked flats, and 3 flats of flats.

Make at least four different numbers in your base, using as few pieces as possible for each.

How can you use what you know about patterns of exponents to create a set of pieces for a different base?

Write a Report About the Different Base

Once you have created and investigated your set of base 3 or base 4 pieces, write a report about your set. Attach to your report one of each kind of piece from the base you chose.

1. Describe the patterns you find in your set, including what your base number to the zero power equals.

2. On a new Stacks and Flats Recording Sheet, write arithmetic expressions for, and record how you made, the numbers 11, 12, 35, and 36.

3. Figure out the next number after 36 that would have a 0 in the units column. Describe how you figured this out.

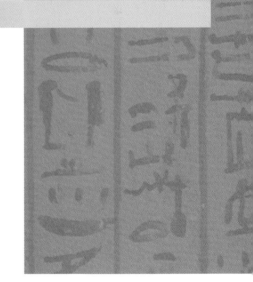

hot **words** | base-two system
binary system

page 88

10 The Power Up Game

Do you think switching the base and exponent will result in the same number? Can a small number with a large exponent be greater than a large number with a small exponent? Play the Power Up game and find out.

Explore Exponents with the Power Up Game

How can you use 3 digits to make the greatest possible expression using exponents?

Play the Power Up game with a partner. Make a chart modeled after the diagram below to record the numbers you roll, the expression you write, what the expression equals, and whether you score a point on the turn.

The Power Up Game Rules

1. Each player rolls a number cube four times and records the digit rolled each time. If the same digit is rolled more than two times, the player rolls again.

2. Each player chooses three of the four digits rolled to fill in the boxes in this arithmetic expression: $(\square + \square)^{\square}$

3. Players then evaluate their expressions. The player whose expression equals the greater number gets one point.

What did you learn about exponents from the Power Up game?

Evaluate Arithmetic Expressions with Exponents

Choose three of the four letters to Dr. Math that you want to answer. Write answers using words, drawings, and arithmetic expressions. Make sure that you describe how you solved each problem. Do not just give the answer.

Can you describe the effect of exponents on numbers in different ways?

Dear Dr. Math,

We're studying exponents in math class. I was asked to draw a picture that showed what 4^3 means. I drew this:

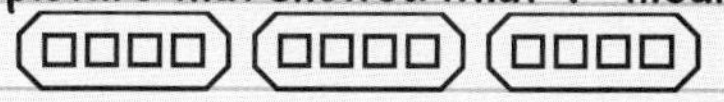

(□□□□) (□□□□) (□□□□)

I drew 3 sets of 4 because 4 gets multiplied 3 times. But I know that $4 \times 4 \times 4 = 64$, so I don't understand why my picture shows 12. Why doesn't it show 64, even though it shows 4^3? Can you explain what's happening with my picture? How would I draw a picture of 4^3?

Exasperated with Exponents

Dear Dr. Math,

Isn't it true that $8 \times 3 = 3 \times 8$? I know I learned this! And aren't exponents a way of showing multiplication? But 8^3 does not $= 3^8$! This is really confusing. Can you tell me why this doesn't work? Is there ever a time when it does work to switch the two digits?

Bamboozled in Boston

Dear Dr. Math,

Something is wrong with my calculator! I think that 3^6 should be much smaller than 5^4, because after all, 3 is smaller than 5. And 6 is not that much bigger than 4. So I don't understand why my calculator tells me that the answer for 3^6 is bigger than the answer for 5^4. I think it needs a new battery; what do you think? Why can't I tell which of two numbers is bigger by comparing the base numbers?

Crummy Calculator

Dear Dr. Math,

One of the problems I had to do for homework last night was to figure out what 10^0 was equal to. I called my friend to ask him, and he thought it was 0. He said it means that 10 is multiplied by itself 0 times, so you have nothing. I thought it was 1, but I don't remember why that works. Which of us is right? And can you please explain to me why?

Zeroing In

hot **words** | exponent
power

Homework
——————
page 89

11 Efficient Number Systems

Some number systems use a base, and others do not.

Here you will decode different place-value systems. You will see that some systems work better than others. Your decoding work will help you think about the features that make different systems work and that make a number system easy to use.

Decode Three Place-Value Systems

How are bases used in a place-value system?

Each of the number systems shown uses a different place-value system. See if you can use the numbers on this page and on the Decoding Chart to decode each system.

1 Figure out what goes in the place-heading boxes (☐) to decode the system.

2 Choose two numbers that are not on the chart. Write an arithmetic expression that shows how the numbers are made in each system. Make sure you label each expression with the name of the system.

Our System	Hand System Place Values				
30		1	0	1	0
35		1	0	2	2
40		1	1	1	1
50		1	2	1	2
60		2	0	2	0
101	1	0	2	0	2

Our System	Crazy Places Place Values					
30			1	0	0	0
32			1	0	0	2
40		1	0	0	0	0
47		1	0	0	0	7
53	1	0	0	0	0	3

Our System	Milo's System Place Values				
30		●	★	★	★
34		●	★	●	●
58		●	◆	◆	◆
105	●	★	★	●	◆

Compare the Features of Many Number Systems

Use the features from class discussion to make a chart that compares some of the systems you have learned in this unit. Include at least one system that is additive, one that uses a base, and one that uses place value.

1 Make a list of at least six different features of a number system.

2 Choose at least six different number systems and describe how they use each feature.

3 Create your own chart format. Leave an empty column so you can add our system to your chart later.

Feature	MD	Chinese Abacus	Milo's System
Place Value	Yes, large beads are worth 10. Small beads = 1.	5 is on top. 1 is on bottom. Yes.	No, because the system use symbols.
base system	Yes, 1, 10, 100.	Base of 5. 5, 50, 500, 5,000 Base of 1. 1, 10, 100, 1,000	Yes, it have base. 1, 3, 10, 30, 100
#s represented in more than one way	Yes, you can use 10 small bead or 1 big bead to make 10.	Yes, you can use 5 ones or 1 5's to make 5.	Yes, you can use U and make 2.

Check Whether Our System Is Efficient

What makes some number systems more efficient than others? Use your chart to check whether our system is efficient or not. Explain your reasoning in writing and use your chart as an example.

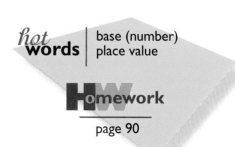

hot **words** | base (number) place value

Homework

page 90

What features make a number system efficient?

A New Number System

You have decoded many different types of number systems and looked at their features. Now it is time to improve one of the systems by bringing together the best of each. You will start by taking a look at the ancient Egyptian system.

Decode the Egyptian Number System

How does the ancient Egyptian system work?

Use what you have learned about decoding systems to figure out the value of each symbol on your Ancient Egyptian System Reference Sheet. On a separate sheet of paper, write the value of each symbol.

1 Choose four numbers that are not on the chart. Write each number in the ancient system.

2 Write arithmetic expressions to show how each of the four numbers is made.

1,024... ...500

Analyze the Ancient Number System

Describe the features of the Egyptian system. What are its disadvantages? Find at least one way to improve the ancient system.

What are the ways in which the ancient system is like our system?

Revise a Number System

Choose one of the following number systems. Find a way to make it more efficient. Present your revised system clearly with words, drawings, and arithmetic expressions, so that others could use it.

What features would a number system need to be efficient?

1 Revise either Alisha's (Lesson 4), Yumi's (Lesson 7), Milo's (Lesson 11), or your own Mystery Device system.

2 Find a more efficient way to make all the numbers between 0 and 120, using words or symbols. Show how at least four numbers are made using arithmetic expressions.

3 Describe how your system uses the following features:

a. place value

b. base system

c. symbols

d. rules

e. a way to show zero

f. trading

g. range (greatest and least number)

h. making a number in more than one way

4 Compare your system to our system and describe the differences and similarities.

Evaluate the Efficiency of an Improved System

A good way to evaluate a number system is to ask questions about the different properties of the system. What are the building blocks of the system? Does it use base or place value? Can a number be made in more than one way? Come up with at least three more questions and use them to explain why your partner's system is or is not efficient. Make sure your explanation talks about the mathematical features of the system.

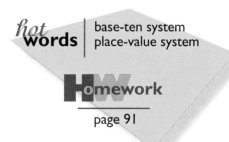

hot **words** | base-ten system
place-value system

Homework

page 91

Inventing a Mystery Device System

 Homework **1**

Applying Skills

1. $3(7) + 5(2) = ?$

2. $6(5) + 3(2) + 7(8) = ?$

3. $4(25) + 6(15) + 2(10) = ?$

4. $9(100) + 4(10) + 4(1) = ?$

5. $5(1,000) + 3(100) + 2(10) + 9(1) = ?$

6. $6(1,000) + 3(10) + 4(1) = ?$

George's Mystery Device System
• Large beads = 10
• Small beads with arms pointing in = 1
• Small beads with arms pointing out = 5

Show each number in George's system. Draw only the beads you need for each number. Remember to draw the diagonal arms either in or out. Use the fewest beads you can.

7. Draw 128 in George's system.

8. Draw 73 in George's system.

9. Draw 13 in George's system.

10. What is this number?

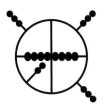

11. What is this number?

12. What is this number?

Extending Concepts

13. What is the greatest number you can make with this system? Explain how you know.

14. Can you find a number you can make in more than one way? Can you find a number that can be made in more than two ways?

15. Is there a number you can make in only one way in George's system? What is the arithmetic expression? What would the arithmetic expression be for that same number written in our system?

Writing

16. Answer the letter to Dr. Math.

Dear Dr. Math,

In my Mystery Device system, for numbers larger than 100, large beads mean 100, small beads with the arms pointing out are 20, and small beads with the arms pointing in are 10. To make numbers less than 100, large beads are 10, small beads with the arms pointing in are 5, and small beads with the arms pointing out are 1. My friends get confused using my system. How should I change it?

B. D. Wrong

Comparing Mystery Device Systems

Applying Skills

Write each number in arithmetic expressions.

1. 6,782

2. 9,015

3. 609

4. 37,126

5. 132,056

6. 905,003

What number do these arithmetic expressions represent?

7. 8(100,000) + 7(10,000) + 6(1,000) + 3(100) + 2(10) + 1(1)

8. 3(10,000) + 1(1,000) + 6(100) + 3(1)

9. 5(100,000) + 2(1,000) + 3(100)

10. 7(1,000,000) + 5(100,000) + 6(10,000) + 2(1,000) + 4(100) + 3(10)

Remember George's system from Lesson 1.

George's Mystery Device System
- Large beads = 10
- Small beads with arms pointing in = 1
- Small beads with arms pointing out = 5

11. Using only 3 beads, draw at least seven numbers in George's system. Only draw the beads that count.

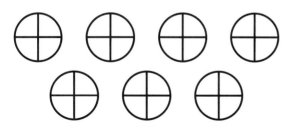

Extending Concepts

12. Using George's system, draw 32 in at least five different ways. Only draw the beads that count.

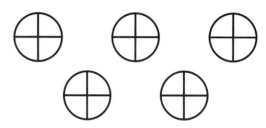

13. Can you think of an explanation for why you can make numbers so many ways in this system but not in our own?

Making Connections

14. Look at the way you solved items **11** and **12.** Did you just keep thinking of different solutions or did you try to use a pattern? Describe the pattern you used or one you might try next time.

Number Words in Many Languages

Applying Skills

Number Words in Fulfulde		
1 go'o	11 sappo e go'o	30 chappan e tati
2 ɗiɗi	12 sappo e ɗiɗi	40 chappan e nayi
3 tati	13 sappo e tati	50 chappan e joyi
4 nayi	14 sappo e nayi	60 chappan e joyi e go'o
5 joyi	15 sappo e joyi	70 chappan e joyi e ɗiɗi
6 joyi e go'o	16 sappo e joyi e go'o	80 chappan e joyi e tati
7 joyi e ɗiɗi	17 sappo e joyi e ɗiɗi	90 chappan e joyi e nayi
8 joyi e tati	18 sappo e joyi e tati	100 teemerre
9 joyi e nayi	19 sappo e joyi e nayi	
10 sappo	20 noogas	

Number	Fulfulde Word	Arithmetic Expression	English Word	Arithmetic Expression
25	a.	b.	c.	d.
34	e.	f.	g.	h.
79	i.	j.	k.	l.
103	m.	n.	o.	p.

1. Copy and complete the chart above.

2. What building blocks does Fulfulde use that English also uses?

3. What building blocks does Fulfulde use that English does not use?

4. Write the Fulfulde number words that match these arithmetic expressions, and tell what each number equals:

   a. $(10)(5 + 3) + 1(1)$

   b. $(10)(4) + 1(5) + 3(1)$

   c. $1(100) + 1(20) + 1(5) + 4(1)$

5. What are the arithmetic expressions for the English number words used in item 4?

Extending Concepts

6. In Fulfulde, the arithmetic expression for chappan e joyi is 5(10). In English, the arithmetic expression for 50 is also 5(10). Find another Fulfulde number word that has the same arithmetic expression as the matching English number word.

Writing

7. How are the Fulfulde number words and arithmetic expressions like the English number words and arithmetic expressions? How are they different?

Examining Alisha's System

Applying Skills

Alisha's Mystery Device System
- Large beads = 4
- Small beads with arms pointing in = 1
- Small beads with arms pointing straight up = 20

Make a chart like the one below and use Alisha's system to show each number on the Mystery Device. Remember to draw in the diagonal arms in the correct position. Next, write the number in Alisha's number language. Then write the arithmetic expression for the number.

Number	Sketch	Number Word	Arithmetic Expression
65	1.	2.	3.
143	4.	5.	6.
180	7.	8.	9.
31	10.	11.	12.

13. How would you commonly write the number word for 36? What is the arithmetic expression for this number?

Alisha's Number Language	
1 en	15 sim-vinta, sim
2 set	16 vinta-vinta
3 sim	17 vinta-vinta, en
4 vinta	18 vinta-vinta, set
5 vintaen	19 vinta-vinta, sim
6 vintaset	20 soma
7 vintasim	30 soma, set-vinta, set
8 set-vinta	40 set-soma
9 set-vinta, en	50 set-soma, set-vinta, set
10 set-vinta, set	60 sim-soma
11 set-vinta, sim	70 sim-soma, set-vinta, set
12 sim-vinta	80 vinta-soma
13 sim-vinta, en	90 vinta-soma, set-vinta, set
14 sim-vinta, set	100 vintaen-soma

Extending Concepts

14. What is the greatest number in the system that you can make in more than one way? Find all the different ways you can make the number and write an arithmetic expression for each one.

15. What is the least number you can make in more than one way? Make a different arithmetic expression for each way to show how the number is made.

 **Homework 5**

Exploring the Chinese Abacus

Applying Skills

Abacus Rules

The beads above the crossbar are worth five times the value of the column if pushed toward the crossbar. Each bead below the crossbar is worth one times the value of the column if pushed toward the crossbar. The value of the column is zero when all beads are pushed away from the crossbar.

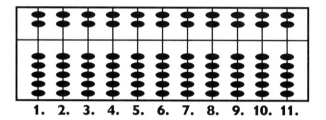

1. 2. 3. 4. 5. 6. 7. 8. 9. 10. 11.

Write the place value using words and numbers for each row marked on the abacus. For example, row 11 = 1 or *the ones place*.

Show each number on the abacus using the fewest beads. Show only the necessary beads. Write the arithmetic notation for each solution.

Number	Sketch	Arithmetic Expression
6,050	12.	13.
28,362	14.	15.
4,035,269	16.	17.

Extending Concepts

18. How is zero shown on the abacus? Why is the zero important? How is zero on the abacus like or unlike zero in our system?

19. Find a number smaller than 100 that you could make on the abacus in more than one way. What is the arithmetic expression for each way you can make the number?

Writing

20. Answer the letter to Dr. Math.

> Dear Dr. Math,
>
> It seems like the Chinese abacus system has two different values for each column. Do the columns have the different values or do the beads?
>
> Out O'Place

How Close Can You Get?

Applying Skills

Show different ways you can make this number on the abacus. Write the arithmetic expression for each solution.

Number	Sketch	Arithmetic Expression
852	1.	2.
852	3.	4.
852	5.	6.

What are some different ways you can make 555? How many beads do you use each time? Make a chart putting the number of beads you used in order from least to greatest.

	Number of Beads Used	Arithmetic Expression
7.		
8.		
9.		
10.		

Extending Concepts

11. What pattern do you see in the number of beads used to make 852 and 555? Why does this pattern work this way? Don't forget to explain how you used trading to make the different numbers.

Writing

12. Answer the letter to Dr. Math.

> Dear Dr. Math,
> To make the number 500, I can use five 100-beads from the bottom, or one 500-bead from on top. Or I can use four 100-beads from the bottom AND two 50-beads from on top. When would it make sense to use a different way to make the number?
> Clu

Additive Systems

Applying Skills

Judy's System Judy invented a new additive system. ◇ = 1 □ = 9 ! = 81		
Judy's System	**Our Number System**	**Arithmetic Expression**
!!□◇◇◇	**1.**	**2.**
!!!!!□□□◇◇◇	**3.**	**4.**
□□□◇◇	**5.**	**6.**
!!!!!!!□□□□◇◇◇◇◇◇◇◇◇◇◇◇	**7.**	**8.**
9.	222	**10.**
11.	98	**12.**

What do these numbers in Judy's system represent in our system? How would you make numbers using Judy's system? Complete the chart and write an arithmetic expression for each number.

Extending Concepts

13. Write 1,776 in Judy's system. What number would you add to Judy's system to make writing larger numbers easier? Make sure your number fits the pattern. Write 1,776 using your added symbol.

14. How did you choose your added number?

Writing

15. Answer the letter to Dr. Math.

> Dear Dr. Math,
> When my teacher asked us to add a new symbol and value to the additive number system we had been using, I added ☆ to represent 0. But when I tried making numbers with it, things didn't turn out the way I planned. When I tried to use it to make the number 90, everyone thought the number was 9. Why didn't people understand? Here's what I did: □☆
>
> Z. Roe

The MD System

Applying Skills

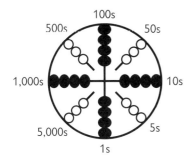

Make the numbers below using the MD system. Use the above illustration of the MD system to help you.

1. 9   **2.** 72   **3.** 665

4. Show 7,957 in the MD system using the fewest beads you can.

5. Make the same number with the most beads you can.

Write the arithmetic notation for how you made each number using the MD system.

6. 9   **7.** 72   **8.** 665

9. 7,957   **10.** 7,957

Extending Concepts

11. What is the greatest number you can represent in the MD system?

12. How do you know there are not any higher numbers?

13. Can you make all the numbers in order up to that number?

14. What is one change you could make to the system, so that you could make some higher numbers?

15. What kind of trades could you make in the MD system?

16. How do you show 0 in the MD system? Is this like having a 0 in our system or is it different? Why?

Writing

17. The MD system and the Abacus system use the same place values. What other similarities do they have? differences? Which do you find easier to use? Explain.

Stacks and Flats

Applying Skills

Make a list of powers through 4 for each number and write the value.

1. $2^0 =$ $2^1 =$ $2^2 =$ $2^3 =$ $2^4 =$

2. $3^0 =$ $3^1 =$ $3^2 =$ $3^3 =$ $3^4 =$

3. $4^0 =$ $4^1 =$ $4^2 =$ $4^3 =$ $4^4 =$

Figure out which base is used in each problem below.

4. $36 = 1(?^3) + 1(?^2)$

5. $58 = 3(?^2) + 2(?^1) + 2(?^0)$

6. $41 = 2(?^4) + 1(?^3) + 1(?^0)$

7. $99 = 1(?^4) + 1(?^2) + 3(?^1)$

Write the arithmetic expression for each number.

8. 25 base 2   **9.** 25 base 3   **10.** 25 base 4   **11.** 78 base 2   **12.** 78 base 3   **13.** 78 base 4

Extending Concepts

14. Describe how you figured out the arithmetic expressions above. Did you use a power greater than 4? Explain why.

15. In the base 2 system, how does the pattern continue after 1, 2, 4, 8, 16, …? How is this pattern different from the pattern 2, 4, 6, 8, 10, 12, …?

16. Look at the patterns in the chart below. Fill in the missing numbers for each pattern. Then answer the three questions at the bottom of the chart.

Number	Patterns of Powers	Patterns of Multiples
2	1, 2, 4, 8, 16, __, __, __	2, 4, 6, 8, 10, 12, __, __, __
3	1, 3, 9, 27, __, __, __	3, 6, 9, 12, 15, 18, __, __, __
4	1, 4, 16, 64, __, __, __	4, 8, 12, 16, 20, __, __, __

    a. How can you use multiplication to explain the patterns of powers?

    b. How can you use addition to explain the patterns of multiples?

    c. Do you have another way to explain either group of patterns?

Power Up Game

Applying Skills

What do these expressions equal?

1. $1(2^4) + 2(2^3) + 2(2^2) + 1(2^1) + 2(2^0)$

2. $2(3^4) + 1(3^3)$

3. $1(4^3) + 2(4^2) + 2(4^0)$

4. $2(3^3) + 2(3^2) + 2(3^1) + 2(3^0)$

5. $2(2^3) + 1(2^1)$

Arrange each set of numbers to make the greatest and least values for each expression. The last blank represents an exponent.

6. 3, 6, 5, 2   greatest (____ + ____)——
            least (____ + ____)——

7. 7, 6, 6, 5   greatest (____ + ____)——
            least (____ + ____)——

8. 5, 2, 3, 4   greatest (____ + ____)——
            least (____ + ____)——

9. 1, 2, 3, 4   greatest (____ + ____)——
            least (____ + ____)——

10. Powers Puzzle Figure out each missing value. The sixteen numbers in the shaded area add to 11,104 when you have finished.

Number	To the 2nd Power	To the 5th Power	To the ___ Power	To the ___ Power
3			27	
	16			
6				1,296
		32		

Extending Concepts

11. What conclusion did you reach about the number that goes in the exponents place when you want a large or a small number? Find 2 digits where the larger digit raised to the smaller digit is bigger than the smaller digit raised to the larger digit. Find 2 digits where the larger digit raised to the smaller digit is equal to the smaller digit raised to the larger digit.

Making Connections

12. Earthquakes are rated on a Richter scale from 1 to 10. They are rated to one decimal place; the most powerful earthquake in North America was in Alaska and was rated 8.5. The power of an earthquake increases 10 times from one whole number to the next. An 8.5 earthquake is 10 times more powerful than a 7.5 earthquake. How many times more powerful is a 6.2 earthquake than a 4.2 earthquake?

Efficient Number Systems

Applying Skills

Show how you would write each of these numbers in these systems.

1. Zany Places

	50	40	20	10	5	1
43						
72						
25						
17						

2. Maria's System ★ = 1 ● = 2 ◆ = 4

	100	20	10	2	1
31					
67					
183					
118					

Write an arithmetic expression for each number.

3. 43, Zany Places

4. 72, Zany Places

5. 25, Zany Places

6. 17, Zany Places

7. 31, Maria's system

8. 67, Maria's system

9. 183, Maria's system

10. 118, Maria's system

Extending Concepts

11. What is the greatest single digit in our base 10 system? What is the greatest single digit in the base 2 system? base 3? base 4? base 9?

12. How does the base of a place-value system affect the number of digits/symbols it contains?

13. What numbers cannot be made in Maria's system? Why?

14. Why can you make any number in Zany Places, but not in Maria's system?

15. Which system do you think is easier for writing numbers, Zany Places or Maria's system? Why?

Making Connections

16. The Metric system uses base 10 for measurement. Write the powers of 10 that are used in the measurements below. How is the English system (inches, feet, yards) different from the Metric system? Can you use exponents to describe the English system?

Metric	Number of Units	Exponents
deka	ten	$10^?$
kilo	thousand	$10^?$
giga	billion	$10^?$

A New Number System

Homework 12

Applying Skills

Give one example of a system you've learned about that uses each property, and explain how it works in that system.

1. place value **2.** base **3.** zero

4. trading **5.** building blocks

George's Mystery Device System

- Large beads = 10
- Small beads with arms pointing in = 1
- Small beads with arms pointing out = 5

George's New and Improved Mystery Device System

George decided to add place value to his system. He made each bead on the horizontal arms worth 100 and each bead on the vertical arms worth 10. He didn't change the value for the smaller beads.

- Large beads on vertical arm = 1 ten
- Large beads on horizontal arm = 1 hundred

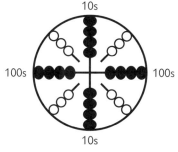

- Small beads with arms pointing in = 1
- Small beads with arms pointing out = 5

Draw each number in George's original system and in his revised system. Beware—some numbers cannot be made.

Original   New and improved

6. 32   **7.** 97   **8.** 156   **9.** 371

Extending Concepts

Write the arithmetic expression for each of the numbers you made in both George's original system and his new and improved system.

10. 32   **11.** 97

12. 156   **13.** 371

14. What is the greatest number that George is able to make in his revised system? Are there any numbers less than this number that George cannot make? Tell why or why not.

15. In what way is George's revised system better than his old system? In what way is George's revised system not as good as his old system?

16. What are some ways that George's revised system is different from our system? Name at least 3 differences.

17. Make a list of all the different properties of our own number system that make it easy to use. For each item on your list, write one sentence explaining why.

Glencoe

The *McGraw-Hill* Companies

This unit of MathScape: Seeing and Thinking Mathematically was developed by the Seeing and Thinking Mathematically project (STM), based at Education Development Center, Inc. (EDC), a non-profit educational research and development organization in Newton, MA. The STM project was supported, in part, by the National Science Foundation Grant No. 9054677. Opinions expressed are those of the authors and not necessarily those of the Foundation.

Send all inquiries to:
Glencoe/McGraw-Hill
8787 Orion Place
Columbus, OH 43240-4027

ISBN: 0-07-866794-1

3 4 5 6 7 8 9 10 058 06 05